TWENTY-TWO MANLY STORIES TO INSPIRE MEN WITH MANLINESS

BOOK ONE

STERLING NIXON

S&J Publishing

The final approval for this literary material is granted by the author.

First printing

All individuals appearing in this work are based on actual manly individuals. Any resemblance to real persons, living or dead, is purely intentional.

ISBN: 978-1-951780-33-3 (Trade paperback)
ISBN: 978-1-951780-32-6 (eBook)

S&J Publishing

Printed in the United States of America

Dedicated to the next Generation of Young Men

A Note to the Reader

Throughout history, acts of courage and self-sacrifice have shaped the destinies of individuals, nations, and civilizations. Some of these moments are well-known, immortalized in legend and history books, while others have faded into obscurity, known only to those who seek them out. This collection of stories is dedicated to the men who stood firm in the face of overwhelming odds—warriors, leaders, and soldiers who chose to fight, not for themselves, but for something greater. Whether they were motivated by loyalty, duty, faith, or sheer defiance, their actions demand recognition.

It is important to remember that each of these stories takes place in a different time and culture, shaped by customs and values that may seem foreign to the modern reader. The men featured in these pages lived in worlds where honor often meant more than life itself, where surrender was considered worse than death, and where war was not just a contest of strength, but of will. Some of the choices they made may seem extreme or even unsettling by today's standards. However, this book does not ask the reader to judge them, nor does it seek to glorify war or violence. Rather, it is an exploration of the raw and unshakable human spirit—the will to stand firm when all others would flee.

Courage takes many forms, and while these accounts focus on battlefield heroics, the essence of bravery is universal. These men did not fight because they believed themselves invincible; they fought knowing they would likely die, yet they chose to stand regardless. In doing so, they transcended their own mortality, becoming legends that endure far beyond their final moments.

Whether it is a samurai holding his ground to protect his lord, a lone warrior refusing to retreat before an army, or a soldier sacrificing himself so that others may live, each of these stories captures the essence of courage. Not all of these men won their battles. Not all of them survived. But they all proved one thing—true bravery is not found in victory, but in the willingness to fight against impossible odds, even when the outcome is already written.

This book is a tribute to those who chose to stand.

The Last Stand of Marcus Cassius Scaeva

Unus Contra Omnes
One Against All

The Roman Republic teetered on the brink of collapse as two of its greatest generals—Julius Caesar and Pompey the Great—clashed for dominance. In the year 48 BC, during the height of the Civil War, Caesar's forces found themselves entrenched at Dyrrhachium, a crucial stronghold on the coast of modern-day Albania. Vastly outnumbered, Caesar had no choice but to construct a massive line of fortifications to encircle Pompey's superior army, hoping to starve his rival into submission. But Pompey had no intention of waiting to die. He launched a counteroffensive, striking at the weakest section of Caesar's fortifications with overwhelming force.

Among the defenders stationed at this critical juncture was a centurion named Marcus Cassius Scaeva, a soldier of unbreakable will and ferocity. A veteran of many campaigns, he had earned his rank through blood and discipline, commanding a cohort of eighty men. As Pompey's warships bombarded the coastline and thousands of legionaries stormed forward, Caesar's lines began to buckle. A section of the defenses collapsed under the relentless assault, and Pompey's forces surged through the breach. The defenders, realizing the danger of being surrounded, began to retreat. But not all of them. One man refused to fall back.

Scaeva stood alone on the last remaining section of the crumbling wall, gripping his sword with iron determination. His men were either dead or fleeing, but he did not move. If the enemy poured through, Caesar's army could be annihilated. The Pompeians, sensing victory, rushed toward him. But instead of finding a broken soldier, they found a whirlwind of death.

The first to approach were cut down in an instant. Scaeva's blade moved with practiced precision, severing limbs and slashing through armor. A spear hurtled toward him, piercing his thigh, but he merely snapped it off and continued fighting. Arrows rained down from above, one embedding itself in his eye. The pain was blinding, but he reached up and tore the shaft free, his face a ruin of blood. He did not falter.

Time blurred as he fought, his shield splintering under the sheer weight of incoming blows. Pompey's soldiers, expecting an easy victory, found themselves facing something out of legend. The lone centurion, battered and broken, held his ground against an entire cohort. He fought with such ferocity that some of his enemies hesitated, whispering among themselves that he must be a favored warrior of the gods. But Scaeva was no divine agent—he was simply too stubborn to die.

The bodies piled around him, a testament to his refusal to yield. He seized fallen weapons from the dead, hurling javelins and daggers at his foes, his arms moving with unrelenting speed. His armor, once gleaming, was now slick with blood—his own and that of the enemy. Hours passed, and still, he stood. The Pompeians, frustrated and unnerved by his resistance, struggled to push forward.

Then, as exhaustion threatened to claim him, reinforcements arrived. Inspired by the sight of a lone man holding back an entire army, fresh legions charged into the breach. With renewed vigor, they drove the Pompeians back, reclaiming the section of the wall that had nearly cost them the battle. The enemy, demoralized and weary, fell back in retreat. When the dust settled, Scaeva still stood—barely.

His body was riddled with wounds, over two hundred in total, yet he remained conscious. His breathing was ragged, his face mangled, his body shaking from blood loss. And yet, his first words were not of pain or surrender, but a simple question—had Caesar won?

Word of Scaeva's stand spread like wildfire through the ranks. When Caesar himself heard of the feat, he could scarcely believe it. The centurion, now half-blind and scarred beyond recognition, was summoned before him. Caesar, ever the pragmatic leader, knew the value of such indomitable will. He personally promoted Scaeva to the rank of primus pilus, the highest honor a centurion could attain. He offered him a fortune in gold, enough to retire as a legend.

But Scaeva refused.

He was a soldier, not a noble, not a merchant. The battlefield was his home, and war was his calling. He would fight again, and again, until Rome no longer needed men like him.

His name became legend, whispered among legionaries who dreamed of achieving even a fraction of his renown. In all of Rome's long history of conquest, few warriors ever matched the sheer unbreakable will of Marcus Cassius Scaeva, the one-eyed centurion who defied death itself.

The Unkneeling Havildar Ishar Singh

ਨਿਬਾਹੀ ਜਾਨੇ, ਲੜੀ ਜਾਨੇ, ਹਾਰੇ ਨੀ

Fight until the last breath, never surrender, never accept defeat

The sun cast a golden glow over the battlefield as a desperate struggle unfolded in the heart of India. It was 1757, and the fate of an empire rested on the shoulders of one man: Havildar Ishar Singh, a warrior of the Sikh Khalsa, who would become legend at the Battle of Saragarhi.

The British Indian frontier was a volatile place, fraught with conflict between the colonial forces and the indigenous tribes who resisted their rule. Along the rugged terrain of the North-West Frontier Province, the British had established a line of defensive forts, small outposts meant to maintain control over the region. Saragarhi was one such post—nothing more than a small signal station manned by twenty-one Sikh soldiers of the 36th Sikhs regiment.

On the morning of September 12, 1897, the ground trembled beneath the advance of an enemy force unlike any seen before. Over 10,000 Pashtun warriors descended upon the outpost, their banners fluttering in the wind, their voices raised in a battle cry that echoed across the hills. The small band of defenders had no illusions about their fate. They were cut off from reinforcements, trapped within their stone walls, with no chance of survival. But they did not waver. They were Sikhs, bound by honor, duty, and an unbreakable will to fight.

At the head of the defense stood Havildar Ishar Singh, a man forged in the crucible of war. His turban was tied tight, his rifle steady in his hands, his eyes locked on the horizon. He knew the battle would end in their deaths. But it was not death that concerned him—it was how they met it. They would not fall as cowards. They would stand as warriors.

The first wave of attackers rushed forward, their curved swords flashing in the sun. The defenders fired in disciplined volleys, cutting down the enemy in droves. The walls of Saragarhi shook under the impact of relentless assaults, but still, they held firm. Ishar Singh moved among his men, rallying them with shouts of encouragement, his sword flashing as he struck down those who made it past the gunfire.

Hours passed, and the walls began to crumble. The enemy, unable to breach the fort through sheer numbers, brought forth explosives. A deafening blast tore through the structure, reducing part of it to rubble. Smoke and dust filled the air as the Pashtuns poured through the gap, their numbers overwhelming the remaining defenders. But still, Ishar Singh fought.

Wounded, his uniform soaked in blood, he refused to yield. His sword, now chipped and stained, struck down foe after foe. When his blade finally shattered, he seized a fallen rifle, wielding it as a club, smashing through the ranks of his enemies. One by one, his men fell, each taking dozens of attackers with them. Yet, none turned their back to the fight.

As the last few defenders stood surrounded, Ishar Singh, with his breath ragged and his body broken, raised his sword one final time. His voice rang out over the chaos, a cry of defiance, a declaration of their unshaken resolve. Then, at last, he fell.

The battle was over. The Pashtuns had taken Saragarhi, but at a terrible cost. Over 600 of their warriors lay dead, slain by just twenty-one men who had refused to surrender. The sacrifice of Ishar Singh and his comrades would not be forgotten. When news of their stand reached the British command, they recognized the battle as one of the greatest last stands in history. Each of the twenty-one soldiers was posthumously awarded the Indian Order of Merit, the highest gallantry award available to native soldiers at the time.

But more than medals, their legacy lived on in the hearts of warriors and patriots. The tale of Saragarhi was passed down through generations, a testament to the unbreakable spirit of the Sikh soldier. In the annals of history, few had stood against such odds. Fewer still had fought so fiercely that their enemies spoke of them with reverence. Havildar Ishar Singh did not seek glory, nor did he fight for empire or wealth. He fought because he was a soldier. He fought because he believed in duty, in honor, and in the bonds of brotherhood. And in the end, he did not fall as a man defeated—he fell as a warrior who refused to kneel.

The Indomitable Spirit of Jan Žižka: The One-Eyed General

Kdož sú Boží bojovníci a zákona jeho!
Ye who are warriors of God and of His law!

The air was thick with the scent of war, the sound of clashing steel and the cries of men echoing through the Bohemian hills. It was the early 15th century, and the Kingdom of Bohemia stood on the brink of chaos. Religious strife had fractured the land, and the Hussite Wars had erupted in the wake of Jan Hus's execution. It was in this turmoil that one man, a warrior unlike any other, rose to defy the might of the Holy Roman Empire and the Papal crusaders—Jan Žižka, the one-eyed general.

A veteran of many battles, Žižka had already lost an eye in his youth, yet he never wavered in his pursuit of victory. His mind, a battlefield of its own, conjured tactics that would redefine medieval warfare. In 1419, he took command of the radical Hussite forces, a ragtag army of peasants and zealots armed with little more than faith and farming tools. Opposed by the armored knights of Europe's most powerful lords, they were expected to be swept aside with ease.

But Žižka saw no defeat, only opportunity. He revolutionized warfare with the wagenburg strategy—fortifying wooden wagons into mobile fortresses, transforming them into deadly bulwarks of resistance. When the crusaders charged, they were met not with cowering farmers but with a wall of wooden carts bristling with spears, crossbows, and early firearms. The invincible knights of Europe crumbled before the ingenuity of a man who saw war not in numbers, but in strategy.

At the Battle of Kutná Hora in 1421, Žižka faced overwhelming odds. The Imperial army, led by the German and Hungarian forces, sought to crush the Hussite rebellion once and for all. Outnumbered, outmatched, and pressed against the frozen Elbe River, Žižka refused to surrender. Instead, he turned the ice itself into his weapon. He lured the crusaders onto the frozen river, where the weight of their armor became their doom. As the ice shattered beneath them, drowning hundreds, Žižka's men descended upon the survivors, leaving none standing.

Victory after victory, Žižka became a nightmare to his enemies. No fortress could withstand his siege tactics, no cavalry charge could break his defensive formations. Even when he lost his remaining eye to a crossbow bolt, plunging him into total blindness, he remained unshaken. Sightless, he still commanded his armies with the same precision, relying on his keen mind and trusted captains to relay the battlefield's movements.

His final campaign came in 1424. Still undefeated, he marched once more against the Imperial forces. Disease, not war, finally felled the warrior who had never known defeat. But even in death, Žižka defied mortality. His last wish was that his skin be flayed from his body and used to make war drums, so that he could continue to lead his men into battle even after death.

His legend endures, a symbol of defiance against tyranny and an undying testament to the power of the human will. To this day, the Czech people remember his legacy, and his battle cry remains etched in history:

Leonidas and the Last Stand at Thermopylae

Μολὼν λαβέ
Come and take them

The sun rose over the narrow pass of Thermopylae, casting long shadows over the assembled warriors. In the distance, the earth trembled under the advance of a colossal army—the vast host of the Persian Empire, commanded by King Xerxes himself. The Great King had come to bring all of Greece to heel, his army numbering in the hundreds of thousands. Against him stood a mere 7,000 Greeks, led by 300 Spartans under the command of King Leonidas.

Leonidas, the warrior-king of Sparta, knew that this battle was no ordinary conflict. The Persian numbers were overwhelming, their might seemingly insurmountable. But to a Spartan, honor meant more than survival. Sparta had been raised on the belief that a warrior's greatest glory was to die defending his homeland. And so, Leonidas and his men prepared to make their stand, knowing they would never return home.

As the Persian army approached, Xerxes sent an envoy, offering the Spartans a chance to surrender and lay down their weapons. He promised riches, power, and their lives—if only they would bow. Leonidas, unwavering, responded with words that would echo through history:

"Come and take them."

The battle began. Persian archers darkened the sky with a hail of arrows, but the Greek shield wall held firm. When the enemy infantry surged forward, they met the unbreakable phalanx of Spartan hoplites, their spears thrusting with deadly precision. Again and again, the Persians charged, and each time they were repelled. The narrow pass nullified their superior numbers, and for two days, the Greeks held their ground, cutting down wave after wave of enemy soldiers.

Xerxes, furious, sent forth his elite troops—the Immortals. Clad in fine armor, wielding deadly swords, they were the best of the Persian army. But they too fell before the discipline and iron will of Leonidas and his men. The Spartans fought like gods, their spears striking true, their shields unyielding. Blood soaked the ground, and the bodies of Persia's finest warriors piled high.

Yet betrayal would seal their fate. A Greek traitor named Ephialtes led the Persians through a hidden mountain path, allowing them to flank the defenders. Leonidas knew the battle was lost, but retreat was never an option. He ordered most of the Greek allies to withdraw, choosing to remain with his 300 Spartans, 700 Thespians, and 400 Thebans to face the final assault.

Surrounded, outnumbered a thousand to one, the Spartans fought to the last breath. Their spears shattered, their swords broken, yet still, they fought with fists, teeth, and sheer defiance. Leonidas fell, pierced by arrows and spears, but his warriors fought on, forming a final shield wall around their fallen king. They perished to the last man, their sacrifice buying Greece the time it needed to unite against Persia.

Though they died, their defiance became legend. Their stand inspired all of Greece, leading to the eventual Persian defeat at the battles of Salamis and Plataea. Even Xerxes himself, surveying the battlefield, could not help but marvel at the courage of his fallen foes. To this day, the words of Leonidas stand as a testament to resistance against tyranny, carved into the stone of Thermopylae's memorial:

Μολὼν λαβέ
Come and take them

Saigō Takamori: The Last Samurai's Defiance

死ぬことより不名誉を恐れよ
Fear dishonor more than death
-Samurai Code of Bushidō

The morning mist rolled over the hills of Shiroyama, veiling the city of Kagoshima in an eerie silence. It was 1877, and Japan was at the crossroads of history. The samurai, once the ruling class of the nation, were being swept aside by the tide of modernization under the Meiji government. Their way of life, their honor, and their code—bushidō—were fading into legend. Yet there was one man who would not go quietly into the past. Saigō Takamori, the last true samurai, had taken up arms in one final stand.

Born into a samurai family, Saigō had once been a key figure in the Meiji Restoration, fighting to overthrow the old shogunate and usher Japan into the modern era. But the new government he helped create had turned its back on the warriors who had built the nation. Samurai privileges were abolished, their swords outlawed, and their very identity erased. Disillusioned, Saigō withdrew to his home in Satsuma, where he trained warriors in the old ways, preparing for a battle he knew would come.

That battle arrived in 1877. The Meiji government, fearing rebellion, sought to crush the last remnants of samurai resistance. Saigō and his followers, vastly outnumbered and outgunned, took up arms in what would become known as the Satsuma Rebellion. For months, they waged a guerrilla war against the imperial forces, their swords clashing against rifles, their honor against the march of time. Victory was never an option. They fought for something greater than conquest—they fought for their dignity.

The final stand came at Shiroyama. Saigō and his remaining 500 samurai faced an imperial army of 30,000, armed with modern artillery and rifles. The night before the battle, knowing death was certain, the samurai gathered for one last feast, speaking of old times, reciting poetry, and sharpening their blades. There was no fear among them—only resolve.

At dawn, the imperial army opened fire, bombarding the samurai position with relentless artillery fire. The mountain shook with each explosion, yet Saigō's warriors held their ground, waiting for the moment to strike. As the imperial troops advanced, the samurai unsheathed their katanas and charged one last time. The battle was swift and brutal. The samurai cut through the ranks of riflemen, but bullets tore through their armor. One by one, they fell, yet none turned to flee. Saigō, wounded but unbowed, fought alongside his men, his blade flashing in the morning light.

As the battle reached its end, Saigō, mortally wounded, was carried to a quiet place by his closest followers. He knew his time had come. Refusing to be captured, he chose seppuku—the ritual suicide of a samurai—to preserve his honor. With his last breath, he died as he had lived, a warrior to the end.

The rebellion was crushed, the samurai extinguished as a ruling class, but their spirit was never forgotten. Saigō Takamori became a legend, honored even by those who defeated him. His defiance symbolized the unwavering heart of the samurai, a reminder of the cost of progress and the price of honor.

His legacy endures in Japan, his spirit immortalized in statues, stories, and in the words that defined the samurai:

死ぬことより不名誉を恐れよ
Fear dishonor more than death

Davy Crockett: The Lion of the Alamo

Be sure you're right, then go ahead

The first light of dawn crept over the battered walls of the Alamo, illuminating the figures of weary defenders standing atop the ramparts. Their rifles rested in calloused hands, their bodies ragged from days of relentless siege. Below them, stretching to the horizon, lay the vast army of General Antonio López de Santa Anna—thousands of disciplined Mexican soldiers ready to storm the walls. Inside the mission-turned-fortress, less than two hundred Texian and Tejano defenders remained. Among them stood a man whose name would echo through history—Davy Crockett, the legendary frontiersman and congressman turned warrior.

Born in 1786 in the untamed backwoods of Tennessee, Crockett was raised in hardship, learning to hunt, track, and fight from a young age. He became a soldier, a scout, and a politician, famed for his sharp wit and frontier spirit. Yet when tyranny threatened Texas, he traded his seat in Congress for a rifle, riding to the Alamo to make his stand.

For thirteen days, the defenders of the Alamo repelled wave after wave of Mexican assaults. Cannons thundered, musket fire cracked, and the walls trembled under constant bombardment. Supplies dwindled, but Crockett and his men did not falter. When Santa Anna's emissary demanded their surrender, their answer was cannon fire. They knew the odds; they knew no reinforcements were coming. Still, they chose to fight.

The final assault came on March 6, 1836. Before dawn, thousands of Mexican troops surged forward in a relentless wave. The defenders fought like demons, reloading between shots with lightning speed, cutting down ranks of soldiers as they climbed the walls. Crockett, wielding his rifle "Old Betsy," fired until his ammunition ran dry. Then he took up his knife and clubbed men with the butt of his weapon, fighting in the thick of the melee. Blood ran in rivers, bodies piled upon bodies, yet the defenders did not retreat.

The battle turned into a brutal hand-to-hand struggle. One by one, the defenders fell, overrun by sheer numbers. Crockett and his last few men made their stand in the chapel, fighting to the very last breath. When the gunsmoke cleared, the Alamo was silent, its walls stained with the sacrifice of those who had chosen to fight for freedom rather than flee.

Santa Anna believed he had crushed the rebellion, but the defiance of the Alamo became a rallying cry across Texas. "Remember the Alamo!" echoed through the ranks of Sam Houston's army, fueling their thirst for vengeance. Just weeks later, Houston's forces routed the Mexican army at San Jacinto, securing Texas' independence.

Davy Crockett's legacy was not one of mere survival, but of an unbreakable spirit. He stood as a symbol of courage against impossible odds, a man who chose to die on his feet rather than live on his knees. His defiant words still inspire those who stand against tyranny:

"Be sure you're right, then go ahead."

The Rise and Fury of Shaka Zulu

Inkunzi isematholeni
The bull is among the calves
– Zulu proverb

The African sun burned high in the sky, casting long shadows over the rolling hills of KwaZulu. The land trembled under the thunderous march of disciplined warriors, their spears glinting in the harsh light. It was the early 19th century, and a new power had risen in southern Africa, transforming a fragmented collection of clans into a single, fearsome kingdom. At the center of this transformation stood one man—Shaka Zulu. His rise was swift, his rule absolute, and his name struck fear into the hearts of his enemies.

Born in 1787, Shaka was the son of Senzangakhona, a Zulu chieftain, and Nandi, a woman of the Langeni clan. His childhood was marred by hardship, as his father cast him and his mother aside, forcing them to seek refuge among rival tribes. But adversity forged him. He learned the ways of war among the Mthethwa, training under Chief Dingiswayo, whose guidance molded him into a warrior of unmatched skill and cunning.

When his father died in 1816, Shaka seized his birthright, uniting the Zulu under his banner. He was no mere chieftain—he was a revolutionary. He transformed the way his people fought, introducing the **iklwa**, a short stabbing spear that turned traditional skirmishes into brutal, close-quarters combat. He reorganized the Zulu military, crafting the deadly buffalo horns formation, where regiments encircled and crushed their foes with unrelenting precision.

Under his rule, the Zulu expanded their domain through warfare and diplomacy alike. He instilled absolute discipline in his warriors, forging them into an unstoppable force. They marched barefoot across the savanna for miles without complaint, their endurance unmatched. They struck with speed and ferocity, overwhelming kingdoms that had stood for generations.

His greatest challenge came in battle against the Ndwandwe, a rival kingdom that had long sought to dominate the region. The two forces met at the Battle of Gqokli Hill in 1818, where Shaka and his vastly outnumbered warriors faced an enemy nearly twice their size. Instead of retreating, he lured them into a trap. His men feigned exhaustion, drawing the Ndwandwe forward, only to unleash a devastating counterattack. The enemy was crushed, their warriors slaughtered, and their power shattered.

Victory after victory followed, and the Zulu kingdom grew. Shaka ruled with an iron will, rewarding loyalty but punishing betrayal without mercy. His warriors adored him, but his enemies feared him even more. Stories spread of his relentless campaigns, his ability to outthink and outfight any who dared oppose him.

Yet power breeds enemies, even from within. As the years passed, Shaka became increasingly harsh, grief over his mother's death twisting his mind. His half-brothers, sensing weakness, struck in **1828**, assassinating him in a coup. As he fell, he is said to have cursed them, warning that his death would bring doom upon the Zulu. And though his killers seized the throne, none could ever match the legacy of the warrior king.

The name of Shaka Zulu lived on, his influence shaping the destiny of his people for generations. His warriors remained among the fiercest in Africa, their tactics studied and feared by colonial powers that would one day seek to claim the land.

Even today, his legend endures, a symbol of strength, strategy, and the power of a single man to change the course of history.

Gaius Mucius Scaevola: The Hand That Defied a King

Fortis Fortuna Adiuvat
Fortune Favors the Brave
-From the Aeneid, Virgil

The city of Rome stood on the brink of annihilation. It was the early 6th century BC, and the fledgling Republic faced a grave threat. Lars Porsena, the mighty Etruscan king of Clusium, had marched on Rome with a vast army, seeking to crush its independence and restore the deposed monarchy of the Tarquins. The city trembled, its defenders struggling to hold the gates, but within its walls, one young Roman refused to bow to fear—Gaius Mucius.

Gaius Mucius was no general, no battle-hardened soldier. He was a youth of noble lineage, but his heart burned with the fire of Rome's defiance. While others debated strategy and defenses, he saw only one path to salvation: striking at the very heart of the enemy. He resolved to infiltrate the Etruscan camp and assassinate King Porsena himself. With nothing but a dagger hidden beneath his tunic, he slipped through the enemy lines, determined to change the course of history. The Etruscan encampment was a sea of warriors, its banners fluttering in the evening breeze. Mucius moved unseen, his eyes fixed on a richly dressed figure seated upon a dais—surely the king himself. Without hesitation, he drew his blade and struck. But fate was cruel. The man he had slain was not Porsena, but his scribe, an official dressed in royal finery. Before he could flee, Mucius was seized by guards and dragged before the true king.

Porsena loomed over the captured Roman, his eyes cold with fury. The punishment for attempted regicide was clear—torture, death, and dishonor. Yet Mucius stood tall, his gaze unflinching. The Etruscans would see no fear in him, no weakness in Rome's sons. When Porsena demanded to know if others conspired against him, Mucius did not beg for mercy. Instead, he thrust his right hand into the flames of a nearby brazier, holding it there as the flesh burned away. He did not cry out, did not flinch. He let his hand burn to ash, his only words spoken through gritted teeth:

"Behold what a Roman can endure!"

Silence fell over the Etruscan camp. Porsena, a man who had conquered kingdoms, had never seen such defiance. This was no ordinary enemy. Rome was no ordinary city. If its people possessed such willpower, its walls would never fall, its people never yield. He released Mucius, now called Scaevola—The Left-Handed, as a sign of respect.

The assassination had failed, but its purpose was fulfilled. The tale of Mucius spread like wildfire, striking fear into the hearts of the Etruscans and emboldening the Romans. Porsena, moved by admiration rather than vengeance, withdrew his army and offered peace. Rome had been saved—not by swords, not by legions, but by the sheer, unbreakable will of one man.

His name became legend, whispered in the forums and inscribed in the annals of history. Gaius Mucius Scaevola had shown the world that Rome would never kneel, that its sons would rather burn than bow. And as long as that fire remained in Roman hearts, no enemy could ever truly conquer them.

The Last Stand of Thorkell the Unyielding

Helst standa en falla á kné
If I must die, let it be standing

The cold wind howled over the fjords, carrying the scent of salt and war. The sea foamed red where the longships landed, their dragon-headed prows now half-submerged in the tide. On the rocky shore stood Thorkell Iron-Eye, his massive frame clad in a wolf-pelt cloak, his axe resting against his shoulder. He had come not to make peace, nor to bargain, but to fight.

Before him, the Saxon army stretched across the valley, banners fluttering in the morning light. Hundreds of shielded warriors, their ranks thick with spears and swords, awaited the clash. Behind Thorkell, fewer than a hundred of his men stood ready—brothers forged in the flames of a hundred battles. The odds did not concern them. Nor did death. They had come for glory, for the songs that would be sung long after their bones had turned to dust.

The Saxon thane stepped forward, his gilded helm gleaming, his voice carrying across the field. "Surrender, Northman. You are outnumbered ten to one. Lay down your arms, and you may yet see another day."

Thorkell smiled, his single eye burning with the fire of battle. "Aye, but each one of us is worth a hundred of you." He hefted his axe, pointed to the horizon where his ships burned, their hulls shattered by Saxon treachery. "There is no home to return to, no retreat to be made. We die here, or we drink from the skulls of our foes!"

The battle began with a thunderous roar. The Vikings surged forward, their shields locking as they crashed into the Saxon line. The impact was like the breaking of a storm against the cliffs, and the Saxons staggered back. Thorkell led the charge, his axe cleaving through helmets and flesh alike. He fought not for land, nor gold, but for the sheer joy of the fight, for the echoes of his ancestors who watched from Valhalla.

The enemy closed around them, wave after wave of Saxons pouring in, hacking and stabbing. One of Thorkell's men fell beside him, his throat cut, but the warrior died with laughter on his lips. "I shall save you a seat at Odin's table!" he called before the light left his eyes.

They fought for what seemed like an eternity, the sun crawling across the sky as the corpses piled high. Blood slicked the ground, making footing treacherous. Thorkell took a sword to the ribs but merely grunted, shattering his attacker's skull with a single blow. He turned to his men—half their number lay dead, but the fire in the eyes of the survivors burned brighter than ever.

The Saxons faltered. Though they had the numbers, they had never seen such fury, such defiance. Some wavered, their confidence shaken. But the thane called for archers, and soon the sky darkened with arrows. The first volley struck deep, shafts piercing through furs and mail. Another fell, then another. Thorkell, breathing heavily, blood pouring from his wounds, stood defiant.

With a roar, he charged one last time, cutting his way through the Saxon ranks, his axe carving a path of carnage. A dozen men fell before him, but he too was slowing, his strength waning. The final blow came swift—a spear thrust deep into his chest, robbing him of breath.

He staggered but did not fall.

With his last strength, he raised his axe high, his voice booming across the battlefield:

"If I must die, let it be standing!"

The Saxons hesitated, watching as the great warrior slowly sank to his knees. Even in death, his grip did not loosen on his weapon.

That night, the ravens feasted well. The Saxons left the battlefield victorious, but the songs of Thorkell Iron-Eye were already being sung in the halls of his people. He had fought as a true Viking, unyielding to the last.

To this day, the warriors of the North remember his final words, carried by the wind across the fjords.

Captain Moroni's Title of Liberty

Title of Liberty
"In memory of our God, our religion, and freedom, and our peace, our wives, and our children."

In 74 BC, the sky burned red with the setting sun as the banners of war fluttered in the warm wind. On one side of the valley stood the Nephite army, their armor dented and their swords bloodied from years of endless war. On the other, the vast Lamanite horde, their warriors clad in hardened leather, their spears and clubs glinting in the twilight. The battle to decide the fate of a nation was about to begin, and at its heart stood one man—Captain Moroni.

Moroni had spent his life defending his people from the Lamanites. He was no king, no ruler by birthright, but a warrior of unmatched brilliance and unwavering faith. He fought not for conquest, but for freedom—the right of his people to live in peace and worship according to their beliefs. Inspired by this battle and later conflicts, Moroni, after seeing the despair creeping into his people as they faced an endless host of Lamanites, took his own cloak and tore it into a banner. Upon it, he inscribed the words:

"In memory of our God, our religion, and freedom, and our peace, our wives, and our children."

Raising the banner high, he rode through the cities, calling every able man to arms. His passion ignited a fire in the hearts of his people, and men flocked to his cause. They would not kneel to oppression. They would not let the Nephite nation fall.

Moroni did not lead his men into battle blindly. He was a strategist, a commander who understood the terrain, the weaknesses of his enemies, and the strength of his own warriors. Knowing that the Lamanites would rely on sheer numbers to overwhelm them, he devised a plan that would turn their aggression against them. He positioned his main force in the valley, making it appear as though they were the sole defenders of the Nephite lands. However, he had secretly stationed Lehi, one of his most trusted captains, with a strong force at a strategic pass behind the Lamanites. As the enemy advanced into the valley, believing they had trapped the Nephites, Lehi's men surged from behind, cutting off their retreat. Surrounded and disoriented, the Lamanites found themselves caught in a deadly vice.

With a cry, the Nephites charged. Shields locked, swords flashing, they met the Lamanites in a thunderous clash. The battlefield became a chaos of steel and blood, men locked in mortal struggle. Moroni was everywhere at once, his sword cutting down foes, his voice urging his warriors forward. The Lamanites fought with savage fury, their sheer numbers threatening to overwhelm the Nephite lines. But as the battle raged, Lehi's force struck from behind, driving panic through the Lamanite ranks. What had seemed like an inevitable victory became a slaughter. The tide had turned.

With the Lamanites now surrounded and outmatched, Moroni ordered his men to cease their attack. He stood before them, a figure of both power and restraint, and called for Zerahemnah to be brought before him. When the Lamanite leader stood in his presence, Moroni offered terms of peace. "Behold, we will end the conflict. If ye will deliver up your weapons and covenant that ye will not return to war against us, we will spare you."

Zerahemnah, humiliated but still filled with rage, responded defiantly. "Behold, here are our weapons of war...but we will not suffer ourselves to take an oath unto you, which we know that we shall break...otherwise we will retain our swords, and we will perish or conquer."

Seeing that Zerahemnah would not accept peace, Moroni returned Zerahemnah's weapons and declared, "Behold, we will end the conflict."

Filled with anger, Zerahemnah retained his sword and rushed forward towards Moroni, but as he raised his blade, one of Moroni's soldiers blocked the weapon, knocking to the ground where it broke by the hilt. With his next strike, the soldier smote Zerahemnah in the head, removing a part of his scalp. The soldier then took the scalp that had fallen to the ground and put it on the point of his sword, saying,

"Even as this scalp has fallen to the earth, which is the scalp of your chief, so shall ye fall to the earth except ye will deliver up your weapons of war and depart with a covenant of peace."

The Lamanites, seeing their leader struck and fearing the fulfillment of the soldier's words, hesitated. Many, realizing the hopelessness of their position, cast their weapons to the ground and made a covenant of peace. However, a few, filled with wrath, refused to surrender and took up their weapons once more. Moroni's forces struck them down, and the remainder of the Lamanite army, terrified and broken, submitted to the Nephites' mercy.

With the battle over, Moroni ensured that the surviving Lamanites, now bound by oath, were allowed to depart in peace. True to his word, he did not seek to destroy them needlessly but upheld the principle of honor and restraint that had defined his leadership. The Nephites, bloodied but victorious, stood upon the field, gazing over the bodies of their fallen brothers.

Moroni, breathing heavily, knelt in the bloodstained sand and whispered a prayer of gratitude. His body was tired, but his spirit burned as brightly as ever. He had lived and fought for one thing above all—the liberty of his people.

The Last Stand of Xiang Yu

生当作人杰，死亦为鬼雄

In life, be a hero among men. In death, be a ghost of legend

The Yangtze River shimmered in the pale moonlight, its waters carrying whispers of fate across the ancient land of China. It was the winter of 202 BC, and the final moments of the great Chu-Han Contention had come. On one side of the river, the remnants of the once-mighty Chu army huddled in silence, their banners tattered, their weapons dulled by endless battle. On the other side stood the victorious forces of Liu Bang, the new emperor of China, whose vast army had encircled them. The Han Dynasty was rising, and the time of the Chu had come to an end. But one man refused to bow—Xiang Yu, the Hegemon-King of Western Chu.

Xiang Yu had been born into war, a warrior of unmatched skill and boundless ambition. He had once stood at the pinnacle of power, leading his armies to shatter the oppressive Qin Dynasty. His strength in battle was the stuff of legend—stories told of him cutting down hundreds of men in a single engagement, his twin swords flashing like the claws of a tiger. His defiance had built an empire, and now his defiance would bring it to ruin.

The Battle of Gaixia had raged for days. Xiang Yu and his remaining 5,000 warriors had fought against an enemy force ten times their number. They had fought not for victory, but for honor. Trapped between the river and the encroaching Han army, they had no illusions about their fate. But they would not go quietly. They would make the earth tremble one last time.

As dawn approached, a cruel trick echoed across the battlefield. From the enemy camp, the songs of Chu could be heard—melodies sung in the dialect of Xiang Yu's homeland. It was psychological warfare at its most ruthless. His men, already weary, faltered at the sound of their people's songs coming from the enemy ranks. The message was clear: their home had fallen, their families had been taken, and there was nothing left to fight for.

Xiang Yu's heart burned with rage, but he knew the battle was lost. He turned to his closest warriors, those who had followed him through countless campaigns, and spoke words they had long feared. "My friends, the heavens have abandoned us. You have followed me loyally, but now I bid you flee. Live, and tell our story."

Many refused to leave his side, swearing to die with their king. With a final charge, Xiang Yu led 800 of his finest warriors into the heart of the Han forces. Like a storm, they crashed into the enemy, hacking through lines of soldiers with terrifying ferocity. The Han warriors, though victorious, quailed before the might of the dying Hegemon-King. Where he rode, men fell like wheat before the scythe. With every stroke of his sword, he carved a bloody path, striking terror even in the heart of Liu Bang's finest generals.

But even legends have their limits. The numbers were too great, the fatigue too deep. His warriors fell one by one, until only a handful remained. Bleeding, his armor shattered, Xiang Yu found himself on the banks of the Wu River. He could have fled, disappeared into the mountains, waged a rebellion in the years to come. But he was no fugitive. He was Xiang Yu, the Hegemon-King. He would not live to see his name tarnished in surrender.

With his final breath, he took his sword and ended his own life, refusing to give his enemies the satisfaction of capturing him. His last words were filled with regret, but also with defiance:

"Heaven has turned against me! I shall not cross this river!"

His death marked the end of an era, but his story did not fade. The people of Chu wept for their fallen king, and even Liu Bang, his greatest enemy, honored his memory. His tale became legend, a symbol of tragic heroism, of courage against impossible odds. To this day, his name is spoken with reverence, a testament to the warrior who would rather die standing than live kneeling.

The Last Charge of Alasdair MacColla

Cha till mi tuille
I shall return no more

The highland mist rolled in thick over the moors, clinging to the rugged hills like a ghostly shroud. In the valley below, the Scottish Royalist forces, battle-worn and outnumbered, braced themselves for the final onslaught. Across the field, the Covenanting army stood in disciplined ranks, their steel-tipped pikes gleaming in the dim light of dawn. The battle for Scotland's soul was reaching its bloody climax, and at its heart stood Alasdair MacColla, the Lion of the Highlands.

Alasdair MacColla was no mere chieftain; he was a warrior whose very name struck fear into the hearts of his enemies. Born into Clan MacDonald, he had grown up in a time of war, his people locked in a brutal struggle against their ancient foes, the Campbells. He had fought in Ireland, learning the ways of Gaelic warfare, leading devastating raids with the ferocity of a man who knew no peace. Now, in the year 1644, he had returned to Scotland, raising the Royalist standard in the name of King Charles, determined to reclaim the Highlands for his kin and his cause.

The battle at Knocknanuss had begun at dawn on November 13, 1647. The Royalists, commanded by James Graham, the Marquis of Montrose, had fought with the desperate fury of men who knew they had no retreat. The charge of the Highlanders had shattered the Covenanting left flank, their claymores cutting a swath of destruction through the enemy ranks. But on the right, where MacColla's men had struck with the weight of thunder, the enemy had regrouped, forming an impenetrable wall of pikes.

MacColla knew the battle was slipping away. His men were exhausted, their ranks thinned by musket fire, their war cries now drowned by the roar of Covenanting cannons. The call to retreat had been sounded, but Alasdair MacColla was not a man to flee. He had never known surrender, and he would not start now.

Raising his sword high, he roared to his men, "For the King, for Clan Donald! Let them know the fury of the Gael!" With that, he spurred his horse forward, leading a final, defiant charge into the heart of the enemy line. The Highlanders followed, their voices rising in a terrible battle cry as they crashed into the Covenanting ranks. The ground shook with the force of their assault, the clash of steel against steel ringing out like thunder across the moor.

MacColla fought like a man possessed, his claymore carving through armor and flesh alike. He was a force of nature, cutting down enemies with the wrath of a warrior who had spent his life in battle. Time and again, he drove his blade into the foe, his men rallying to his side, refusing to yield. But the Covenanting ranks held firm, their pikes bristling like the thorns of a monstrous beast. One by one, the Highlanders fell, overwhelmed by sheer numbers and relentless musket fire.

Surrounded, blood streaming from a dozen wounds, MacColla fought on. His breath was ragged, his limbs heavy, but still, he would not fall. The enemy closed in, a ring of steel tightening around him. A voice called for his surrender, but he only laughed—a deep, fearless laugh that carried over the battlefield. "Come and take me, if ye dare!" he bellowed, his sword still raised, his spirit unbroken.

Then, in a flash of steel, the enemy struck the final blow. Alasdair MacColla, the scourge of the Campbells, the terror of the Lowlands, fell upon the bloodied earth, his claymore still gripped in his dying hands. The Covenanting forces, though victorious, could not deny the honor of the man who had stood against them till the very end.

His death marked the end of an era, but his legend lived on. The Highlanders whispered his name in reverence, the bards sang of his deeds, and his enemies, even in victory, spoke of him with awe. To this day, his tale is told as one of the fiercest warriors Scotland ever knew, a man who met death not with fear, but with defiance in his heart and steel in his hands.

The Defiance of Pharaoh Ahmose I

N ḏd nḫt ḥr ḫm.t ḫswt, n.t ḥ□ n Kmt ḫpr m m□□t m snf

"We shall not bow to the foreign yoke, for the land of Kemet is ours by right and by blood."

The sands of Egypt were stained with blood as the sun set over the city of Avaris. It was the year 1550 BC, and for nearly a century, the Hyksos had ruled over Lower Egypt, their foreign kings sitting upon the throne that rightfully belonged to the sons of Ra. But now, the tides of war had shifted, and the last battle for Egypt's freedom was at hand. Standing at the head of his army, his war chariot dusted with sand, was Pharaoh Ahmose I, the man who would reclaim his kingdom or perish in the attempt.

Ahmose had been only a child when his father, Pharaoh Seqenenre Tao, fell in battle against the Hyksos. His brother, Kamose, had taken up the mantle of war, leading campaigns deep into the enemy's territory, striking at their strongholds with fire and fury. But Kamose, too, had been lost to the struggle, leaving Ahmose as the last hope of the Theban line. For years, he had trained, hardened his heart, and forged alliances, preparing for the day when he would drive the invaders from his homeland.

The battle for Avaris was to be the final reckoning. Ahmose had spent years weakening the Hyksos, seizing their outposts, cutting off their supply lines, and breaking their allies. Now, the enemy was trapped in their last bastion, and the time had come to end their dominion. The Theban army, reinforced with warriors from Nubia and the lands of Upper Egypt, surrounded the walls of Avaris, their siege engines pounding at the gates. Ahmose himself led the charge, his war axe gleaming in the fading sunlight.

The Hyksos, desperate and cornered, fought with the fury of men who knew they had no escape. Their chariots clashed against the Egyptian lines, their archers raining down volleys from the walls. The battle raged for hours, the streets of Avaris turning into rivers of death. Ahmose, mounted on his chariot, cut through enemy ranks with relentless precision, his warriors rallying around him. Every strike of his blade was vengeance for his father, for his brother, for the countless Egyptians who had suffered under foreign rule.

At the height of the battle, Ahmose led a final assault on the palace of the Hyksos king, Khamudi. With a roar, his warriors stormed the stronghold, cutting down the last defenders. Ahmose himself struck the killing blow, driving his spear through Khamudi's chest, ending the rule of the Hyksos once and for all. As the sun dipped below the horizon, Avaris fell silent, and with it, a century of foreign oppression was washed away in blood and fire.

With the Hyksos defeated, Ahmose did not rest. He pursued the remnants of their forces beyond Egypt's borders, ensuring they would never return. His victory not only reclaimed the throne of Egypt but marked the beginning of the New Kingdom, an era of unparalleled power and prosperity. His defiance, his unyielding will, had restored Egypt's glory and set the foundation for an empire that would endure for centuries.

The warriors of Egypt spoke his name with reverence, and the scribes etched his deeds into the temple walls, ensuring that none would forget the king who had freed his people. Even today, Ahmose I is remembered as the great liberator, the pharaoh who cast off the chains of oppression and forged a new destiny for his land.

Hernán Cortés and the Burning of the Ships

Conquistar o perecer
Conquer, or perish

The waves crashed against the shore as the wind carried the scent of salt and conquest. It was the year 1519, and the Spanish expedition, led by Hernán Cortés, had landed on the mysterious shores of the Aztec Empire. Behind them lay the vast Atlantic Ocean, an expanse that separated them from the safety of Spain. Before them loomed the unknown, a land filled with dangers, riches, and warriors who would rather die than surrender their kingdom. It was in this moment, standing on the beaches of Veracruz, that Cortés made a decision that would change history—he burned his ships.

Cortés had arrived with just over 600 men, a small force compared to the massive armies of the Aztec ruler, Montezuma. His expedition had been fraught with challenges, from political disputes in Spain to distrust among his own men. Some of his officers, fearing the overwhelming odds, whispered of retreat. They longed for the safety of Cuba, the Spanish stronghold in the Caribbean, and viewed this campaign as doomed before it had truly begun.

But Cortés would not allow hesitation to take root. He knew that in order to conquer the mighty Aztecs, his men needed to be committed beyond doubt, beyond fear. There could be no turning back. So he gave the order that would forever seal their fate—he commanded his ships to be scuttled and burned, destroying any hope of return. The flames rose into the sky, the wooden hulls cracking and splitting apart as smoke billowed toward the heavens. The ocean swallowed what remained, leaving only the land before them. The message was clear:

Victory or death.

The men stood in stunned silence, watching as their last ties to the old world were consumed by fire and water. Some were furious, others terrified. But slowly, understanding dawned upon them. There was no escape, no retreat—only the path forward. With their leader standing unshaken before them, they took up their arms with renewed resolve. They would not return to Spain as failures; they would carve their names into the pages of history.

With no way to flee, Cortés led his men inland. Through cunning diplomacy and ruthless warfare, he turned the enemies of the Aztecs into his allies. The Tlaxcalans, long oppressed by Montezuma, joined the Spaniards, swelling their ranks. Battle after battle, siege after siege, the forces of Cortés pushed forward. They fought with muskets, swords, and steel-clad determination, cutting a path through the mighty empire of Tenochtitlán.

For two years, the war raged. The Aztecs, fierce and unyielding, met the Spaniards in brutal combat. Cortés and his men endured ambushes, starvation, and sickness, but their resolve never wavered. The burning of the ships had bound them to their mission with an iron will. By 1521, after months of siege and bloodshed, the great city of Tenochtitlán fell. The Aztec Empire, once one of the most powerful civilizations in the world, lay in ruins.

Cortés had gambled everything, and he had won. His name became legend, spoken in fear and awe. The destruction of his own fleet had not been an act of madness, but of absolute conviction. It was the ultimate test of his men's courage, the final push that turned an uncertain expedition into an unstoppable force. History would remember him as the man who burned his way into the heart of an empire and emerged victorious.

Tecumseh and the Last War for Freedom

Neshna chaya ne liwiyo thitewa nole nenewe nepemi neya wilawekawa

Live your life so that the fear of death can never enter your heart

The wind howled through the great forests of the Ohio Valley as warriors gathered beneath the darkened sky. It was the year 1811, and the land that had once belonged to their ancestors for generations was now threatened by the relentless expansion of the American settlers. At the center of the council fire stood Tecumseh, the great Shawnee leader, his piercing gaze sweeping over his assembled warriors. His people were at a crossroads—fight for their lands and way of life, or watch everything they had ever known be consumed by foreign rule. For Tecumseh, there was only one choice.

Born in 1768, Tecumseh had spent his entire life witnessing the steady encroachment of white settlers onto Native lands. He had seen treaties broken, villages burned, and entire tribes forced from their homes. He knew that no single tribe could stand against the might of the United States, but united, the native nations could be unstoppable. With this vision, he traveled far and wide, forging alliances between the Shawnee, the Creek, the Cherokee, the Ottawa, and many others. He preached resistance, not just for land, but for dignity, for the survival of their people.

By 1811, his confederation had grown strong, but danger loomed. Governor William Henry Harrison of the Indiana Territory saw Tecumseh as a threat and sought to break his alliance before it could become a true force. Tecumseh knew that war was coming. But before he could gather more warriors, Harrison struck first, sending his forces to attack Prophetstown, the spiritual center of the resistance, while Tecumseh was away rallying more allies. On November 7, 1811, at the Battle of Tippecanoe, the Americans clashed with Tecumseh's brother, Tenskwatawa, and his warriors. The battle ended in destruction, the village burned, and the confederation shaken but not broken.

When Tecumseh returned and saw the ruins of Prophetstown, his heart burned with rage. He knew there could be no more hesitation. His warriors, disheartened, looked to him for leadership, for reassurance that their cause was not lost. He turned to them and declared that there would be no retreat, no surrender. "The Great Spirit has given us this land, and we will not leave it!" he cried. He commanded his warriors to stand firm, to prepare for battle, and to abandon the ways of appeasement. The time for diplomacy had ended. He burned what little remained of Prophetstown himself, leaving nothing for the enemy to reclaim. From that moment on, they would fight with nothing behind them—only forward, only toward victory or death.

With no home left to return to, Tecumseh and his warriors joined the British in the War of 1812, knowing that this was their last chance to drive back the Americans. He led his forces with unmatched brilliance, winning battle after battle against the U.S. Army. His warriors fought like ghosts in the forests, striking with speed and disappearing into the trees. His reputation grew, and even his enemies spoke of his courage and tactical genius.

But fate had already turned against him. In October 1813, at the Battle of the Thames in present-day Ontario, Tecumseh and his warriors made their final stand. Outnumbered and abandoned by their British allies, the warriors of the confederation stood alone. Tecumseh, knowing that defeat was imminent, did not waver. He fought until the last, leading his men in a desperate charge against the enemy. The battle was fierce, but the Americans, led by William Henry Harrison, overwhelmed them. Tecumseh fell that day, his body never recovered, his spirit never broken.

Though his life ended on the battlefield, his dream did not die. His name became legend, a symbol of defiance, unity, and the relentless fight for freedom. The tribes he sought to unite never regained their lands, but his vision and his courage inspired generations to come. He had burned his past behind him, leaving no path but that of resistance. His people would never forget him, nor would history.

Scipio Africanus: Conquer or Die

Aut vincere, aut mori
Conquer or Die

The winds howled across the African coast as the Roman fleet arrived at the shores of Utica in 204 BC. The Second Punic War had ravaged the Mediterranean for over a decade, and now, the fate of Rome itself rested on the shoulders of one man—Publius Cornelius Scipio, later known as Scipio Africanus. After years of devastation at the hands of Hannibal in Italy, Scipio had done what no Roman before him dared: he had taken the war to Carthage's doorstep.

The Senate had hesitated to approve his campaign, doubting that Rome had the strength to launch an invasion while Hannibal still roamed Italy. But Scipio knew that the only way to end the war was to strike at the very heart of Carthage. He had crossed the sea with a force of hardened legionaries, their shields scarred from past battles, their spirits lifted by the chance to finally bring the fight to their enemy's homeland.

Upon landing, some of his officers questioned the wisdom of such a bold move. They reminded him that if they failed, there would be no retreat. The Carthaginian navy still ruled parts of the sea, and a prolonged siege could cut them off from reinforcements. But Scipio had already made his decision. That night, he gave the order that would forever bind his men to their cause. He commanded the fleet to be burned.

As the flames consumed the Roman ships, the soldiers stood in stunned silence, watching as their only route home turned to ash and embers. Some murmured in fear, realizing there was no turning back. But then Scipio, standing tall before them, raised his sword high and spoke words that would echo through the ages. "We did not come here to return to Rome! We came here to claim victory! Our only road home lies through Carthage!" His voice carried across the shore, and as the fire crackled behind him, the hesitation in his men's hearts melted away. They were not invaders with a chance to retreat. They were conquerors, with no option but to win.

With their path set, Scipio led his army inland. The Carthaginians, realizing the Romans had committed themselves fully, scrambled to muster their defenses. For months, Scipio outmaneuvered them, defeating their forces in open battle and forging alliances with local Numidian leaders, including the great cavalry commander Masinissa. His tactics, inspired by the very strategies Hannibal had used against Rome, broke the Carthaginian resolve piece by piece. By 202 BC, he met Hannibal himself at Zama in the climactic battle that would decide the war.

The battle was fierce, the earth shaking beneath the charge of Carthaginian war elephants. But Scipio had prepared his men, ordering them to create gaps in their lines to allow the beasts to pass harmlessly through. As the elephants lost their momentum, the Roman cavalry, led by Masinissa, struck the Carthaginian flanks with devastating precision. The legions surged forward, their swords cutting through Hannibal's ranks, driving the once-invincible army of Carthage into chaos. By day's end, the great general of Carthage had been defeated, and the Second Punic War was over.

Scipio returned to Rome, not by sea, but through history itself. He was given the name Africanus, the conqueror of Carthage, the savior of Rome. His decision to burn the ships had not been madness—it had been genius. By removing the temptation of retreat, he had forged an army of warriors who fought as if their very survival depended on victory, because it did.

His name would be remembered for centuries, not just as a tactician, but as a leader who understood the power of commitment. For when faced with an uncertain path, he did not hesitate. He burned the way back, ensuring there was only one direction left to march—forward, into legend.

The Last Stand of the Swiss Guard

Fide et Virtute
By Faith and Valor
-Motto of the Swiss Guard

The bells of St. Peter's Basilica tolled in the distance as smoke and screams filled the streets of Rome. It was May 6, 1527—the day the Eternal City fell to fire and blood. The walls of Vatican City trembled as 20,000 soldiers of the Holy Roman Emperor Charles V stormed its gates, their mercenaries bent on plunder, destruction, and slaughter. Standing between them and the Pope was a force of just 189 men, the Pope's most loyal protectors—the Swiss Guard.

For years, the Swiss Guard had been the last line of defense for the papacy. Their reputation as elite warriors stretched across Europe, their discipline forged in the brutal wars of the Swiss cantons. They had sworn an oath not just to defend the Pope, but to give their lives for him if necessary. Now, on that fateful morning, their oath would be tested like never before.

The attack came suddenly, a storm of steel and fury. The German Landsknechts, Spanish Tercios, and Italian mercenaries of Charles V swarmed into Rome, pillaging as they went. The Swiss Guard, vastly outnumbered, formed a defensive line at the steps of St. Peter's Basilica. Their mission was clear: defend Pope Clement VII at all costs, hold back the invaders long enough for him to escape.

The first waves of attackers crashed against the Swiss formation, but the guards held firm, their halberds cleaving through the ranks of the invaders. The streets ran red with blood as they fought with unmatched ferocity, their training and discipline keeping them standing against impossible odds. The roar of cannon fire and the clash of swords echoed through the holy city. Still, they did not yield.

As the battle raged, Pope Clement VII, dressed in simple white robes, was hurried through a secret passageway—the Passetto di Borgo—by a small contingent of Swiss Guards. Their captain, Kaspar Röist, remained behind with the rest, knowing there would be no salvation for those who stayed. Their task was not to survive, but to buy the Pope enough time to flee.

The final stand took place on the steps of the basilica. Surrounded, cut off, and exhausted, the last of the Swiss Guard formed a shrinking circle, fighting back with halberds, swords, and daggers. The enemy demanded their surrender, but there was none to give. Röist himself was struck down, his men cut down one by one. Yet they fought to the last man, refusing to abandon their sworn duty. By the time the battle ended, 147 of the 189 Swiss Guards had been slain.

Their sacrifice was not in vain. Clement VII reached safety in Castel Sant'Angelo, saved by the bravery of his protectors. The sack of Rome continued for weeks, a brutal testament to the chaos of the age, but the legacy of the Swiss Guard was sealed in blood and honor.

To this day, the Swiss Guard continues to stand watch over the Vatican, their ranks filled with men who carry the same unbreakable spirit. Each year, on May 6, new recruits swear the same oath, standing upon the graves of those who gave their lives in 1527.

The Revenge of the Forty-Seven Rōnin

Chūgi no tame ni shinu koto wa, bushi no nasake nari
To die for loyalty is the way of the warrior
- *Bushidō, the samurai code of honor*

The snow fell in thick, silent drifts upon the streets of Edo. The night was cold, but the hearts of the men gathered in the shadows burned with fire. It was December 14, 1702, and for nearly two years, these men—now masterless samurai, rōnin—had waited, planned, and sacrificed for this single moment. Before them lay the heavily guarded mansion of Kira Yoshinaka, the man responsible for the death of their beloved lord, Asano Naganori. Tonight, there would be no retreat, no hesitation. Tonight, they would restore their master's honor or perish in the attempt.

Two years earlier, in 1701, Lord Asano had been humiliated and provoked by Kira, a corrupt court official. Forced to draw his sword within Edo Castle, an unforgivable crime, Asano was ordered to commit seppuku—ritual suicide—without being given the chance to defend himself. His retainers, now rōnin, were left without a master, dishonored and cast into disgrace. But their loyalty did not waver. Led by Ōishi Yoshio, they swore an oath of vengeance, knowing that to act against Kira meant certain death.

For nearly two years, they scattered, concealing their intentions. Some disguised themselves as merchants, others as monks. Ōishi himself feigned a life of drunken debauchery to convince Kira's spies that he had abandoned all thoughts of revenge. But in secret, they sharpened their blades and waited for the perfect time to strike. When Kira's vigilance had waned, when he least expected it, they would come.

And so, on that bitter winter night, the Forty-Seven Rōnin moved as one. Dividing into two groups, they attacked from the front and rear, scaling the walls of Kira's mansion. The guards, caught off guard in the dead of night, were no match for the disciplined warriors. The clash of steel echoed through the corridors as the rōnin cut down their enemies with relentless precision. Every strike was for their master. Every life taken was payment for the crime that had stolen his.

Kira's men fought desperately, but the outcome was never in doubt. The rōnin moved with the grace and fury of men who had already accepted death. Room by room, they advanced, cutting down resistance until they reached Kira himself. The once-arrogant court official cowered in a storeroom, trembling before the warriors who had come to deliver justice.

Ōishi stepped forward, his sword steady. "You are granted the same honor our master was denied," he declared, presenting Kira with a blade to take his own life. But Kira, dishonorable to the end, refused. Without hesitation, Ōishi struck him down and severed his head, the final act of vengeance completed.

The rōnin did not flee. They did not hide. As dawn broke over Edo, they marched solemnly to the grave of Lord Asano at Sengaku-ji Temple, laying Kira's head before their master's tomb. They had restored his honor. Their duty was complete.

When the shogun learned of their actions, he faced a dilemma. They had broken the law, yet they had upheld the highest ideals of bushidō—loyalty, duty, and honor. The verdict was delivered: the Forty-Seven Rōnin were to die, but not as criminals. They were granted the right to commit seppuku, an honorable death befitting samurai.

On February 4, 1703, one by one, they took their own lives in the same dignity with which they had avenged their lord. They died as warriors, forever etching their names into the soul of Japan.

Even today, their graves at Sengaku-ji are a place of reverence, a testament to the unbreakable bond between a lord and his loyal retainers. Their story is not just one of vengeance, but of duty beyond death, a reminder that true honor is worth any price.

The Last Stand of Francisco de Sá Carneiro at the Siege of Diu

Morrer com honra é viver para sempre
To die with honor is to live forever

The salty winds of the Arabian Sea whipped against the fortress walls as cannon fire thundered across the battlements. It was February 1546, and the city of Diu, a vital Portuguese stronghold on the coast of India, lay under siege. For months, the combined forces of the Gujarat Sultanate and the mighty Ottoman Empire had battered its walls, determined to expel the Portuguese from their lucrative trading outpost. Inside, the defenders—starving, exhausted, outnumbered—knew there would be no reinforcements. If Diu fell, so too would Portugal's control over the Indian Ocean. But one man refused to surrender. Francisco de Sá Carneiro, a Portuguese knight, stood at the heart of the defense, willing to fight to the very last.

The siege had begun the previous year, with the Sultan of Gujarat seeking vengeance for past Portuguese victories. With Ottoman backing, he launched a campaign to destroy the Portuguese fort. Thousands of soldiers, war elephants, and siege engines pressed against the city's defenses, while Turkish artillery pounded the walls. The defenders, numbering barely a few hundred, fought with the ferocity of men who had no choice but to hold their ground. Supplies dwindled. Men fell. Yet, Carneiro led charge after charge, refusing to let despair take root.

As the months dragged on, disease and hunger ravaged the defenders. They fought on empty stomachs, their wounds festering under the relentless sun. Carneiro, wounded multiple times, never wavered. He moved from the walls to the barracks, urging his men to hold firm. He reminded them that to surrender meant death—not just for them, but for Portugal's future in the East. To fall now would be

to yield the seas to their enemies forever.
By early February, the outer defenses had collapsed, and the enemy surged into the streets. The last of the Portuguese defenders, fewer than a hundred men, fell back to the inner citadel. With their backs to the sea, Carneiro stood among them, his sword red with the blood of those who had dared breach the fort. He had one last message for his men: "No retreat. No surrender. If we are to die, let us die as warriors, and let our names live forever."

The final attack came at dawn. The enemy, believing victory was at hand, poured into the citadel in waves. But the Portuguese met them with a fury no one had expected. They fought like demons, every strike fueled by sheer defiance. Even as their numbers dwindled to a handful, even as the walls collapsed around them, they did not yield. Francisco de Sá Carneiro was one of the last to fall, his body surrounded by those of the enemy. But the courage of the defenders had bought time. Reinforcements, seeing the impossible stand at Diu, arrived days later to drive the invaders from the ruins of the fortress. The Portuguese had held their grip on India, and their dominion over the seas remained unbroken.

Though Carneiro and his men perished, their legacy endured. The Siege of Diu became a symbol of undying resolve, a tale whispered by sailors and warriors alike. It was not the might of an empire that had saved Portugal's fate in the East—it was the unbreakable will of a few men who refused to bow, even in the face of death.

The Unbreakable Will of John Chapman

These things we do that others may live
-The motto of the U.S. Air Force Pararescue (PJ) teams

The freezing winds howled across the barren peaks of Afghanistan as gunfire echoed through the valley below. It was March 4, 2002, in the midst of Operation Anaconda, and U.S. Air Force Combat Controller John Chapman was about to make the ultimate sacrifice. High above the valley of Takur Ghar, a Chinook helicopter carrying Navy SEALs and Chapman had come under heavy enemy fire. A rocket-propelled grenade struck the aircraft, forcing an emergency retreat—but not before one of their men, Petty Officer Neil Roberts, was thrown from the ramp onto the snowy mountainside below.

Without hesitation, Chapman and the SEAL team made the decision that would change everything: they would return to rescue their fallen brother.

As the rescue team touched down under relentless enemy fire, Chapman led the charge up the steep, icy slope toward an enemy bunker. With his rifle raised, he engaged the enemy at close range, eliminating multiple fighters and clearing a path for his team. The battle was chaos—bullets snapped through the air, grenades sent shrapnel tearing through the snow, and the sheer altitude left the soldiers gasping for breath.

Then, as Chapman advanced toward another enemy position, a burst of machine gun fire struck him, knocking him to the ground. His teammates, believing him mortally wounded, were forced to withdraw under overwhelming firepower. The helicopter carrying the retreating SEALs took off, leaving Chapman alone on the mountain.

But he was not dead.

Badly wounded, alone, and surrounded by enemy fighters, Chapman regained consciousness. Rather than hide or surrender, he picked up his rifle and resumed the fight. He crawled through the snow, dragging his broken body to a better position, and continued engaging the enemy. He fought off wave after wave of advancing fighters, holding his ground against impossible odds. When an enemy fighter moved in to execute an incoming U.S. rescue team, Chapman forced himself to his feet, exposing himself to deadly fire to provide cover.

Wounded multiple times, he fired his last rounds, protecting the very men who had thought him dead. He stood alone against an overwhelming enemy force until he fell, his final act one of unyielding courage. When U.S. forces finally retook the mountain, they found Chapman's body surrounded by the lifeless forms of enemy fighters. He had fought to the last breath, never relenting, never surrendering.

Sixteen years after his death, in 2018, John Chapman was posthumously awarded the Medal of Honor, making him the first Air Force service member to receive the nation's highest military decoration since the Vietnam War. His heroism was later confirmed through drone footage, showing his last stand in stunning detail—an unbreakable warrior who refused to die without a fight.

Chapman's actions on Takur Ghar serve as a testament to the warrior spirit, a reminder that true heroism is not in surviving, but in refusing to let others fall, no matter the cost.

The Last Stand of Torii Mototada

武士の名誉は長生きすることではなく、立派に死ぬことにある
A warrior's honor is not found in living long, but in dying well
-Samurai code of Bushidō

The rising sun cast long shadows over the castle walls of Fushimi. It was August 1600, and Japan stood on the brink of the greatest war in its history. Two mighty factions—the forces of Tokugawa Ieyasu and those loyal to Ishida Mitsunari—prepared for a final reckoning. But before the decisive Battle of Sekigahara, one man would make the ultimate sacrifice to ensure his lord's victory. That man was Torii Mototada, a samurai whose loyalty would be tested beyond measure.

Mototada had served Tokugawa Ieyasu since childhood. He was no mere retainer—he was a brother-in-arms, a trusted warrior who had fought in countless battles. When Ieyasu marched his forces westward to meet Mitsunari, Mototada was given a grim duty: hold Fushimi Castle at all costs. It was a near-suicidal command. The castle, though strong, was isolated, and Mitsunari's army numbered over 40,000 men. Mototada had fewer than 2,000.

He did not hesitate.

With his warriors, he reinforced the walls, readied his archers, and prepared for a siege that could only end in death. Before the first attack, he gathered his men and spoke with unwavering resolve. "A warrior's honor is not found in living long, but in dying well," he declared. "We do not fight for survival. We fight so that our lord may stand upon Sekigahara unburdened. Let the enemy come. We shall show them the meaning of samurai steel."

The assault began with a relentless barrage of arrows and gunfire. The sky darkened with smoke as siege engines hammered the castle walls. Wave after wave of enemy soldiers stormed the gates, only to be cut down by the fierce defenders. Mototada fought alongside his men, his katana slicing through the ranks of enemy samurai. Blood soaked the stone pathways, but still, they held firm.

For ten days, the siege raged. When the outer defenses crumbled, Mototada and his remaining warriors retreated deeper into the castle, fighting room by room, hallway by hallway. Food and water ran low, but their spirits did not waver. They fought not for victory, but for honor, for duty, for the legacy of the Tokugawa.

On the final day, with the enemy breaking through the last stronghold, Mototada stood in the courtyard, his armor dented, his blade dripping with the blood of his foes. He had one last duty to fulfill. He knelt on the bloodied stones and prepared for seppuku, the ritual death of a samurai. With unwavering hands, he took the blade and ended his life in the manner of the greatest warriors before him.
Fushimi Castle fell, but its defenders had bought what they had promised—time. The sacrifice of Mototada and his men delayed Mitsunari's forces, allowing Tokugawa Ieyasu to regroup and march to Sekigahara. There, the fate of Japan was decided, and the Tokugawa rose victorious, establishing a shogunate that would rule for over 250 years.

Torii Mototada's name lived on, not as a man who sought survival, but as a samurai who embraced death for the sake of duty. His last stand remains a testament to the unbreakable will of the warrior class, a reminder that true loyalty is measured not in words, but in sacrifice.

The Defiance of Cynaegirus

Κρεῖττον ἐστὶν ἐλεύθερον ἀποθανεῖν ἢ δοῦλον ζῆν.

Better to die free than live as a slave

-Spartacus, 73 BC

The crash of waves against the shores of Marathon was drowned out by the roar of battle. The year was 490 BC, and the Persian Empire had come to subjugate Greece. They had landed in great numbers, their forces stretching across the fields like an unbreakable tide. But against them stood the Athenians and their allies, a fraction of their size, yet unyielding in their resolve. Among them was Cynaegirus, a warrior whose name would become legend before the day was through.

As the Persian army formed its ranks, the Athenians charged, their phalanx crashing into the enemy lines with the force of a thunderbolt. The fighting was brutal, spears shattering, swords flashing, men locked in the deadly struggle of shield against shield. Though outnumbered, the Greeks fought with the desperation of free men unwilling to kneel before a foreign king. Slowly, the Persian line buckled, and then, in a moment of chaos, it broke.

The invaders turned to flee, retreating toward their ships. The Athenians, sensing victory, pursued them across the bloody sands, cutting them down as they ran. But the Persians were not beaten yet. If they could escape and regroup, they could sail to Athens and take the city while its defenders were exhausted from battle. The Athenians knew they had to stop them from reaching their ships.

Cynaegirus, a warrior of great renown, was at the forefront of the chase. He saw the enemy scrambling onto their ships, seeking safety beyond the waves. He would not allow it. With a roar, he surged ahead of his comrades, racing past fleeing Persians, dodging arrows and spears. He reached the shoreline just as one of the great Persian warships was pushing off from the beach. Without hesitation, he leaped onto its side, gripping the rail with one mighty hand. With his other, he swung his sword, cutting down the sailors who tried to push him away. He clung to the ship with the strength of a man possessed, determined to drag it back to shore by sheer force of will.

The Persians, desperate to rid themselves of this mad warrior, struck at him with blades and axes, but still, he held on. In their panic, one of them raised a sword and brought it down upon his arm, severing it at the shoulder. Blood sprayed across the deck, but Cynaegirus did not scream. He did not let go. With his remaining hand, he reached up and grasped the ship anew, refusing to surrender.

Again, the Persians struck. His other arm was hewn from his body, yet still, he tried to hold on, biting into the wood of the ship, his final act one of sheer defiance. At last, overcome by wounds, he fell into the waves, his lifeblood mingling with the sea, his name immortalized in the memory of his people.

Though he perished, his sacrifice was not in vain. The Athenians, inspired by his courage, pressed forward, overwhelming the remaining Persians and securing victory at Marathon. The great invasion was stopped, and Greece remained free—for now.

Cynaegirus did not live to see his homeland safe, but he had done what few men could. He had fought until his body gave out, refusing to surrender even in the face of certain death. His story, told for generations, became the very essence of Greek heroism.

Other Books Written by Sterling Nixon

Historical:
Twenty Manly Stories to Inspire Men with Manliness (Book One)

Historical Fiction:
Gladiators of the Naumachia

Dystopia Fiction:
Acadia
Titan
Charron

Young Adult Fiction:
Nickle Brickle'Bee: In the Heart of EarthWorks
Nickle Brickle'Bee: In the Halls of Harbordeen
Nickle Brickle'Bee: In the Home of Atlantia
Nickle Brickle'Bee: In the Floating Isles of Balinbar
Nickle Brickle'Bee: In the Throne Room of StormHaven
Nickle Brickle'Bee: In the Caves of the Goblin King
Nickle Brickle'Bee: In the Streets of EarthWorks

Made in the USA
Monee, IL
04 November 2026

67387896R00036